Cultivating Magic Mushrooms

W.A. Arbelo

Contents

Page 3 ....Introductory notes

Page 4 ....Materials needed

Page 5 ....Basics of sterile technique and theory of Cultivating

Page 8 ....Outline of Technique

Page 8 ....Spore syringes

Page 10 ....Preparing Jars and Tubs

Page 15........................Hydrating Rye Berries

Page 17........................Sterilizing Jars

Page 17 .... Inoculating Jars

Page 19 ....Colonizing Jars

Page 19 .... Preparing substrate for tubs

Page 20 .... Adding Substrate to tubs

Page 21 ....Colonizing tubs

Page 21 ....Introducing Fruiting Conditions

Page 21 .... Maintaining fruiting conditions

Page 22 ....Harvest

Page 23..........................Drying and storing

Page 24..........................Rehydrating substrate

Page 25..........................Index

Introductory Notes

First off, let me explain exactly what I will be teaching you in this guide. I will be focusing on the specific species of fungi, Psilocybe Cubensis. This mushroom is illegal in many countries due to its content of psilocybin; a strong hallucinogenic substance used recreationally spiritually by many. This being said, this guide is purely for informational purposes and I am in no way responsible for what you use it for. I am going to go over my method for mass producing magic mushrooms. I will go over two different sizes; the smaller can yield you about two ounces of dried mushrooms per flush. (I will go over what a flush is later in the book). The larger will yield much, much more, 4-6 oz per flush.

Most forum sites have some techniques which leave out some major details that many seasoned cultivators assume are basic knowledge, and they will have you making fatal (to your crop) errors early in the process. Many people on forums advise beginners to start with a method called the 'PF tek'. I advise against the "PF tek". Its yields are small and it is prone to contamination.

There are methods using agar and petri dishes which will increase your yield, however this is getting more in depth into microbiology and I want to get straight to growing and harvesting pounds of mushrooms fast for beginners or someone who wants to move from "PF tek" to a bulk grow. We will be covering Multispore Inoculation.

Be ready to invest about $500 for start-up costs for all materials. Follow all the steps in this guide to the T and you will have yourself a bulk harvest in no time.

Materials Needed

Inorganic materials

- **Pressure cooker. All american 921. $$$**
- 3 ML syringes with 18 gauge needles
- Black Trash Bags
- 66 QT Clear Plastic Tub For Large grow preferably with clear lid
- 20 QT Clear plastic Tub for Small Grow preferably with clear lid
- LIghting Timer
- Clamp on work light or sunlight
- 5,000-6,500K CFL light bulb
- Cling wrap
- Ball Wide Mouth Quart Jars
- 2in hole saw drill bit
- Drill
- Paper towels 2 rolls to be safe
- Countertop
- Nitrile/rubber Gloves
- Large Pot that's not your pressure cooker
- Polyester pillow filling/craft store fake snow
- Lighter
- Half inch drill bit able to drill through Ball Jar Lids
- Digital humidity detector (not needed but helps)
- 5 Gallon Bucket with screw on lid attachment
- Sporicidin/Bleach

- QT jar lids with inoculation port (optional but will increase sterile teqnique
- Shot Glass
- Spray Bottle
- Scalpel
- Resealable sandwich baggies
- Oscillating fan or preferably an air purifier with Hepa filter.

Organic Materials

- P. Cubensis Mushroom Spores
- Whole Rye berries in bulk
- Coco coir Brick
- Vermiculite
- Gypsum (not Necessary but it helps)
- Bottled water (reverse Osmosis)

Sterile Technique and Basic process of Cultivation

The most important thing that I can't stress enough is sterile technique. In this chapter I will cover the basic process we will undergo to cultivate P.cubensis mushrooms and how sterile technique is the most crucial part to growing a lot of healthy mushrooms vs coming out empty handed.

Don't think of mushrooms as a plant as they are far from it. Mushrooms are a fungus, and a fungus is a complex organism which will only produce fruiting bodies (mushrooms) in the most optimal conditions. Mushrooms are the reproductive structures created by the **Mycelium** which grows underground much like a trees roots. In the wild you may see a few cubensis mushrooms growing in a cluster because only that small area of the mycelium had optimal fruiting conditions. Our objective is to create the largest most sterile environment for the mycelium to flourish and fruit from everywhere on the mycelium substrate. Which is only possible without getting any Contaminates/Germs in your jars or tubs until the mycelium has grown throughout the entire container.

The process of the mycelium spreading over the substrate, in this case rye berries, is called **colonizing**. Once mycelium has 100% colonized it's container it is much more resistant to being contaminated and can fight off any forein invaders and the forein invaders (mold, bacteria, ect) have no free berries to colonize themselves.

We will first grow our mycelium in QT jars on rye berries. The berries in the Jars will eventually be completely covered with pure white "Mold" (this is the mycelium of P.Cubenis not some ordinary mold). **If any of the jars have any other mold in them or odd slimy patches they must be discarded**; this mold will cause the P. Cubensis mycelium to compete hindering its ability to produce mushrooms. Inspect your jars thoroughly for contaminates. It can take up to a couple months for your jars to completely colonize depending on the strain. Keep your jars in complete darkness and don't knock or move them to avoid damaging the fragile mycelium, unless you are inspecting for contaminates

The next step will be to create a new substrate in our tub. The theory is, you pasteurize your new substrate to kill off most contaminates. You then empty your 100% colonized QT jars into the new substrate and break all of the berries apart and thoroughly mix the berries into the substrate. Now it's the colonizing waiting game again. Leave your tubs in a dark room to colonize. Periodically check for contaminants on the surface of your substrate. Anything growing that is not pure white mycelium is bad. Sometimes a small orange secretion is ok this is a natural antibiotic produced by the mycelium. However a large amount of it indicates large bacterial contamination. The tub can be fruited separately from any other tubs however your yield will just be lower than an uncontaminated tub. If your substrate has any green black or brown mold, any mold other than pure white mycelium it should be discarded so spores will not contaminate your fruiting room/area.

The next stage is the one you'll be ecstatic to have made it to without any contaminations, the Fruiting stage.

As I said earlier, mushrooms are not plants, Plants breathe in CO_2 and expel O_2 or oxygen. Mushroom mycelium is the complete opposite; mycelium breathes oxygen and expels CO2, much like mammals. When your jars are colonizing the mycelium is in a high CO2 environment, constantly expanding in search of the oxygen it needs to fruit it's mushrooms.

    Mushrooms will fruit in complete darkness but your yield will be smaller. Optimal conditions for your mycelium to fruit will be a rapid switch from high CO2 to high oxygen content and the introduction of light.

    Once switched to fruiting conditions mushrooms will start to appear. These are known as pins. Once your substrate has many pins that about a half inch tall it will take about a week for them to develop into full grown mushrooms. Once you harvest your mushrooms more will grow back when the substrate is reintroduced to fruiting conditions, this is called a flush. I will cover more in depth when to exactly to harvest and how to get the most out of multiple flushes in the final chapter of the guide.

    You must have a clean environment to work in when making your spore syringes and when transferring your grains to your tubs. Wear nitrile gloves when breaking up your rye berries and mixing into the substrate. Wear a dust mask to avoid coughing or sneezing on your work area and contaminating your substrates. If you have access to a lot of money you can invest in a flowhood to work under; this is unnecessary but will increase your odds of success even more. I usually work in a room without any airflow. Turn off the central air and fans and wait a while for all dust to settle, as what we call 'dust' can contain anything from bacteria to mold spores. Dust is a contaminate. We only want sterilized substrate and mycelium in our jars and pasteurized substrate and mycelium in our tubs, nothing else.

    On the following page I will breakdown the steps we will take to go from spore to harvest.

Outline of technique

The basics of the technique are as follows:

1. Make spore syringes
2. Prepare Tubs and jars
3. Hydrate Rye Berries
4. Sterilize Jars
5. Inoculate jars with spore syringes
6. Colonize Jars (May take 2 months depending on strain)
7. Once colonized Prepare Substrate for Tubs
8. Add colonized rye berries to tubs
9. Colonize tubs (may take 1 month+)
10. Introduce fruiting conditions
11. Maintain fruiting conditions
12. Harvest
13. Drying and storing your harvest
14. Rehydrate substrate for more flushes
15. Repeat harvest and rehydration for a total of 3 flushes

Making spore syringes

You will need:

- 3ml syringes
- Reverse osmosis bottled water
- Spore print
- Scalpel
- Shot glass
- Lighter

The first thing you will need to do is acquire a P. cubensis spore print. A spore print is a sterile piece of paper with mushroom spores on it. A cultivator cuts the cap off of a fully matured mushroom and lays it gill side down on a sterile piece of paper. A glass is put over top to keep moisture in. Over the next 12-24 hours a significant amount of spores will fall from the cap onto the piece of paper. And you have a spore print!

Choosing a strain

Different strains of P. Cubensis have different qualities; some are more resistant to contamination while others are less resistant some are more potent than others. Some people will tell you they all have the same psychoactive effects, but I notice definite differences in the effects of different strains. For example A+ Albinos were very difficult to grow without contamination. They took a very long time to colonize but were very potent and had very uplifting Giggling effects. The slow colonization could also be due to genetics, however from experience I would recommend B+ or golden teacher strains for first time growers. They are much more resistant to contamination and colonize much faster.

Now to make your spore syringe make sure you are in a room with no moving air where dust has settled. Unfold your spore print and have a scalpel ready, you will scrape some of the spores into your shot glass. The amount of spores does not have to be exact as there are millions in just a small pile of dust. You will probably get 3-6 3ml syringes out of an average sized print.

Once you scrape some spores into your shot glass, take the cap off of your syringe and fill it with some bottled water. Squirt the water into the shot glass and mix the spores around. Suck all of the water and spores into the syringe and replace the cap. You should see some clumps floating in the syringe. Repeat with new syringes until your spore print is gone.

Preparing jars and tubs

What you will need:

- 4 - 8 Wide Mouth Quart mason Jars
- ½ in Drill bit able to drill through Jar Lid
- 2in Hole saw Drill bit
- Drill
- Polyester pillow stuffing / craft store "fake snow" or halloween cobwebs.
- Bleach or sporicidin
- Cling Wrap

To prepare our jars we need to make a hole in the lid to allow us to inject our spores into the jar and to allow the exchange of gases. The mycelium needs oxygen, but having only a small hole will keep the jar at a high CO2 to O2 ratio. This will make sure that the mycelium expands throughout the jar as fast as possible.

The amount of jars you will inoculate will depend on the size tub you are going to be fruiting in. The more jars you put in will usually mean more mushrooms you will get out to an extent. For our smaller Tub the 20qt you only need 2 colonized jars mixed with substrate. For our large grow 4-6 colonized jars is optimal. However you should make 8-10 jars if you want to use 6 in your tub to compensate for contaminated jars. Leftover uncontaminated jars won't go to waste, you can always make a small tub grow out of the extra jars if you don't have any that get contaminated.

Step 1 - Drill one hole with the 1/2in bit in the center of every jar lid.

Step 2 - Pull a tuft of polyester fibers through the hole you drilled, so that the loose ends stick out the top of the lid and the inside is pulled tightly with no fibers hanging into the jar. The polyester should be tight enough to keep gnats from easily getting in but loose enough to allow gasses to freely pass through.

Step 3 - Clean the jars throughly of metal flakes. Use bleach or sporicidin to sterilize the inside of jars. Coat the inside of the jar and let sit for 5 min. Then rinse very thoroughly a few times. Any residue left will kill your mycelium and ruin your jar.

** If you wish you can find online, plastic jar lids made specifically for inoculating substrates. Sold by mycology supply stores, they have a gas transfer patch and a self closing reusable inoculation port. They will go through the pressure cooker a few times before the patch falls off; I've had less contaminations using them vs the pillow stuffing method. The lids cost about $6 each so if going for cost and growing bulk the pillow stuffing is much more affordable. If you have 10 large tubs going at once you'll need about 80 jars sterilized at once, so at $6 per lid x 80jars is $480 just in lids. For getting technique down with one tub or a small tub it's worth it but for a large grow it's not worth it.

For your Tubs you will need your 2in hole saw bit and your polyester stuffing.

You can experiment with different hole placement to see what works for you. Personally I like **3 holes on each side of the tub**. 2 lower and the middle higher. see example 1 and 2 . Too many holes and your substrate will dry out and with too few holes your mushrooms will be starved of fresh air. You just want to make sure the bottom of the lowest hole is about 3in from the bottom of the Tub to leave room for your substrate.

You will eventually stuff polyester stuffing through the holes but for now we are going to take our cling wrap and wrap it tightly around the sides of the tub. You want to go around the sides so you can still open the lid. If the cling wrap doesn't seal well you can seal it with a strip of tape. We want there to be holes in the tub for later but we will need our tub airtight for our substrate to colonize.

Examples of fresh air hole placement

Example 1.

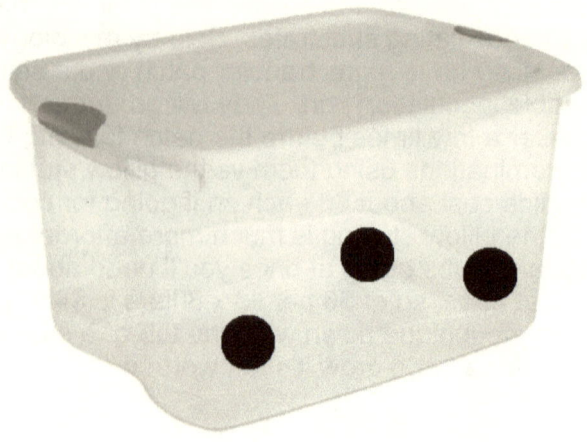

Example 2

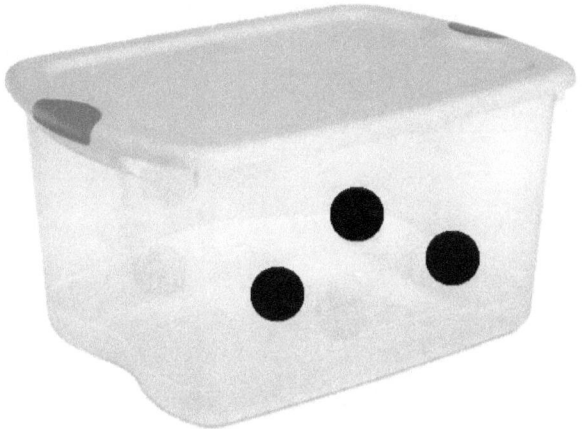

Example 3

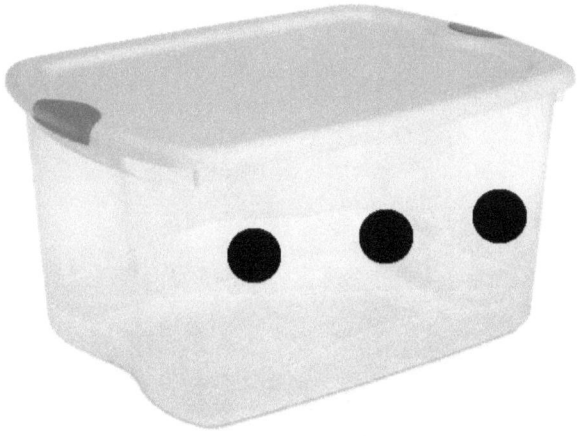

Example 4

Example 5

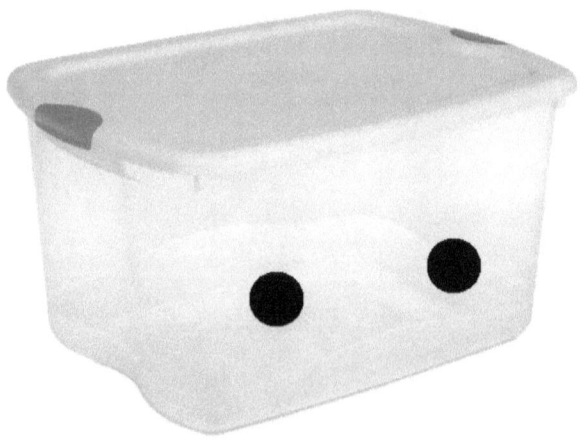

Hydrating Rye Berries

You will need

- Rye berries
- Gypsum
- Tap water
- Large pot
- Stove
- Strainer
- A lot of Paper Towels
- Your jars with holes stuffing in lids

    I've found the best place to get rye berries from is your local natural foods store. Ask if they can bulk order you a bag of **whole** rye berries. Specify whole otherwise you may get crushed berries. If they ask what you need them for just tell them you use them in recipes because they're so healthy. I paid about $40 for a 25lb bag and took a week to get it.

    Once you have you're berries you need to measure them out. Take a quart jar and measure about a half of a jar or a little more dry berries. This amount will equal a full jar of hydrated berries. If you are going to make 8 jars put 8 half jars into the pot. You can pre hydrate berries and fill jars if you plan on sterilizing and inoculating them with spores within the next day or two but preferably immediately.

    Dump the berries straight into the pot. Once you have your 8+ jars worth of berries in the pot you need to rinse them thoroughly. Cover them with tap water and stir them around with your hand. Sterile technique is not important right now because we are going to sterilize the berries in our pressure cooker. Dump the water out of the pot. You may want to use a strainer so you don't lose berries down your drain.

Repeat this until the water is clear when you swish the berries around. About 3 rinses should suffice. **Do not skip this step! Rinsing is very important to avoid contamination!**

Next, fill the pot with twice as much tap water as berries. If the berries take up ¼ of the pot fill the water up to ½ way.

Add 1-2 tablespoons of gypsum and stir it around. Cover the pot let sit overnight, about 12 hours.

Next put your pot with water and berries straight on the stove and bring to a rolling boil. Keep an eye on the berries. You want each berry to be full of water but not exploded into mush. If a couple explode they're done. With practice you will get them perfectly hydrated without any exploding.

While the pot is coming to a boil roll out an entire roll of paper towels on a countertop a few layers thick. Cover the entire countertop. The Idea is to strain the berries and quickly pour them and spread them out on the towels so that the water on the outside of the berries evaporates off as steam. Once evenly spread into a thin layer let them sit. Move them around a little if the layer is pretty thick. Once the steam stops they should be ready to put into your clean jars. If you lightly press your palm onto the berrys and they don't stick to your palm they are ready. The goal here is to have the outside of the berry dry and the inside of the berry moist, the optimal conditions for our mycelium.

Fill the jars ¾ of the way full with your now hydrated berries. Before putting the lid on take a clean paper towel and wipe the glass of the inside sidewall of the jar between the top of the berries and the rim of the jar to remove any residue, water or particles of berries. You want this area of the jar clean and dry. Screw on your lid with the hole we drilled earlier and repeat for the rest of the berries.

Sterilizing jars

Fill your pressure cooker with your jars standing upright. I fit two extra laying down on top of the jars standing up. Put your bobber on the pressure release valve on the 15-PSI mark. Pressure cook on 15-PSI for 90 minutes. Make sure to watch the pressure gauge and lower the stove temperature if the pressure builds too much.

After 90 minutes remove the bobber to release the pressure.

Remove your jars and dry off the outside.

You will have to let them sit until they reach room temperature. You can clean a meat thermometer with rubbing alcohol and check the temperature through the polyester filled hole in the lid if you are in a rush. However, this will increase your risk of contamination in that jar. I recommend letting the jars cool overnight.

Inoculating Jars

You will need:

- Spore syringe
- Lighter
- Paper towel
- Rubbing alcohol

Now that we have our jars with 100% sterilized rye berries, we have the perfect conditions to introduce our P.Cubensis spores. The idea here is that the P. Cubensis mycelium will grow throughout the entire container without any other mold or bacteria growing.

With that being said, now that we have a sterile environment inside of our jars you will want to take your sterile technique very seriously. Anything in the air or on your hands that gets on the needle of the syringe could contaminate the grains within your jar.

On the next page I will break down the steps to inoculate your jar.

Our first step is to decontaminate the needle of our syringe.

1. Wet a paper towel with rubbing alcohol
2. Shake your syringe to break up any clumps of spores and distribute them evenly through the solution
3. Remove the cap from your syringe's needle
4. Wipe the needle with your alcohol soaked paper towel
5. Heat the needle with your lighter until it is red hot glowing orange
6. Squirt a little bit of the solution onto another paper towel to cool the needle. It will sizzle and when it stops it's cool. A small squirt will suffice.
7. Stick your needle through the polyester filled hole in the lid of the jar. Angle it towards the side of the jar and squirt 1-ML onto glass in one spot so it drips down the inside sidewall of the jar. This will allow us to track the growth of our mycelium because the spores will germinate near the side of the jar.
8. Repeat steps 2-7 for each jar you inoculate.
9. Set the jars in a dark cupboard around 70-75/ room temperature. Hotter will allow contaminates to grow faster. While cooler will slow the growth of our mycelium.

Colonizing Jars

Check your jars daily or weekly for contaminations, but be gentle with them any knocking will disrupt your mycelium and slow its growth.

Once the jars are about 30% colonized we want to shake them. You want to break apart all the mycelium that has grown into a clump and disperse it evenly throughout the entire jar. This will cause the mycelium to begin growing on all of the grains which were previously too far to reach. This will speed up your colonizing time frame exponentially. You may have to bang the jars on a bicycle tire or carpeted floor, but be careful or you may shatter a jar. Set the jars back in a cool dry place and leave unto the mycelium has recovered and completely covers all of the rye berries with thick white mycelium.

Check for contaminants; anything thats not bright white mycelium is a contaminant and the jar will have to be disposed of.

Preparing Substrate for Tubs

You will need:

- One brick coco coir per 66qt tub
- 1 qt Vermiculite per brick of coco coir
- Tap Water
- Large pot
- 5 gallon bucket with screw on lid attachment
- Large clean spoon cleaned with rubbing alcohol or soap and water

1. You will need to drill a 1/2in hole in your screw on lid for your 5 gal bucket and stuff it with some polyester fibers. This will allow for steam pressure to release.
2. Place your brick of coco coir in your 5 gallon bucket.
3. Add your Vermiculite

4. Bring 1 - 1.5 gallons of water to a boil
5. Add the water to your 5 gallon bucket
6. Screw the lid on and let sit for 1 - 2 hours.
7. Open your bucket and stir the mixture to evenly distribute your vermiculite throughout the substrate.
8. You now have pasteurized substrate!

Adding Rye and Substrate to Tub

You will need:

- Pasteurized substrate
- 100% colonized rye Berries in qt jars
- Rubber gloves
- Clean room with still air
- Black trash bags
- Bleach solution

1. First we will prepare our tub; spray the inside of the tub with bleach solution and wipe down with paper towels. Be careful not to leave any bleach residue.
2. Next cut and fold your black trash bag to line the base of your tub. You want the bag to cover the bottom of the tub and up the sides 2-3 inches. This will keep mushrooms from fruiting on the bottom of your tub; they will grow anywhere there is light and moisture. The trash bag will hopefully make the conditions on the top of your substrate the most optimal for fruiting.
3. You will want to wipe the trash bag clean with bleach solution as well and allow to dry. From this point you will want the lid on your tub unless you are mixing your substrate.
4. Next empty 6 quart jars of colonized rye berries into the tub. For a smaller tub size use 2-4 jars.
5. Empty your pasteurized coco coir mixture into the tub on top of the rye berries. Enough to make the entire mixture 2-3 inches thick.

6. With Sterile Nitrile/rubber gloves on use your hands to thoroughly mix the berries evenly through the coco coir mixture and level the top as flat as possible.
7. Put the lid on the tub and place the tub where it will be in the dark for the entire colonizing time.
8. You can check the tub after a week or two but it may take up to a month or two to completely colonize.

Colonizing Tubs

Once your substrate is completely covered with solid white mycelium your tubs are 100% colonized.

Look at it very closely you will see tiny white heads forming these white heads are the earliest form of mushrooms! Leave your tub in the dark until these tiny white heads cover the top of your substrate.

Now it is time to move your tubs into fruiting conditions!

Introduce Fruiting Conditions

You will need:

- Polyester fibers
- Fan
- Humidity detector
- Work light or clamp light with 5000k - 6500k bulb or sunlight

Now that you are seeing tiny pinheads all over the top of your substrate you will introduce fruiting conditions to your tub.

Remove the cling wrap from the sides of your tub and stuff the holes with polyester fibers. Pull the loose fibers through to the outside of the tub so no loose fibers hang into the inside of the tub. You want the holes stuffed tight enough so that air can pass through but tight enough to keep dust and insects out.

You should get a humidity detector with a detachable probe. Hang the probe through one of the holes in the tub.

Hang a clamp light over your tub and set a timer for 12 hours on 12 hours off. If you want you can use residual sunlight from a window but optimal fruiting conditions are a 12/12 light cycle.

Maintaining Fruiting Conditions

To maintain fruiting conditions you want to maintain your humidity, light schedule and fresh air exchange into your tub.

You want to keep your humidity at 100% if it keeps dropping low you will want to stuff your polyester fibers tighter to hold in more moisture. If it drops too low you can use a spray bottle to mist the inside of the tub. However with your holes in the right placement and stuffed the right amount you should never have to mist your tub.

Keep an oscillating fan on low speed in the room on 24/7. If you have access to one run a HEPA air filter 24/7 as well.

As your mushrooms develop they will change from tiny whte pinheads into tiny mushroom heads. Once they are about a half inch tall it will be about a week until they are developed enough for harvest.

You may notice some tiny mushrooms die and turn black and never grow larger. These are called aborts. They are more potent per weight than full grown mushroom so don't discard them, they are not contaminated.

Check for contaminations daily; any mold growth other than white mycelium is contaminants. You can put a salt water paste on green mold to slow it down but at that point your crop will already be compromised and smaller.

Harvest

You will need

- Nitrile/Rubber Gloves
- Trays to put your harvested fruits on

Once your mushrooms are full size, the caps will start to open. If you want the most potent shrooms harvest them right before the caps open. If you want the most weight let the caps open but try to harvest before the spores drop, or your mushrooms will look black and dirty.

1. To harvest turn off your fans in your room and let the air settle.
2. With gloves on grip the base of the mushroom or clusters of mushrooms and break the mushroom off of the substrate. Try to tear up as little of the substrate as possible.
3. Place your mushrooms on a large tray so you can easily move them around if needed.
4. Once all the mushrooms are harvested replace the lid on your tub. We will come back to it later.

Drying and Storing

To begin drying you will want to, immediately after harvest, put a fan on high directly on your trays of mushrooms. It may take a couple of days, but in a dry climate your mushrooms should completely dry this way.

You will then want to make what is known as a dessicant chamber.

You will need:

- DampRid
- short plastic tubs
- paper towels or some sort of chicken wire or screen.

Fill your short tubs with about an inch of Damp-Rid brand desiccant. **Make sure to get unscented.**

Lay a layer of paper towels or screening over you desiccant. You don't want your mushrooms to touch this stuff, it will liquify and soak into them.

Put your mushrooms on the screen or paper towels with the desiccant underneath and put the lid on the container. Do not let the mushrooms touch eachother. After a day or two in the chamber your mushrooms will be completely dry and then you can store them in a jar with a silica gel packet, or a layer of Damp-Rid underneath a paper towel in the bottom of the jar. ** I using this storage method be sure to keep the jars upright to avoid mushrooms coming into contact with the DampRid.

I am in no way affiliated with DampRid or their parent company WM Barr. I mention their product only because it is a superior Desiccant and use it for a purpose unintended by the company.

Rehydrating Substrate

What you will need:

- Tap water

- Drain

Now that you've harvested one flush it's time to get the most out of your Substrate. A substrate will continue to produce fruits as long as the conditions are optimal and there are still nutrients in the substrate. You should be able to get three flushes from your substrate.

To rehydrate your substrate you want to soak it with water. Tap water is fine because at this point your substrate being fully colonized is very resistant to contamination; it has a strong immune system.

For your second flush you will be able to just soak the substrate with a spray bottle once and maintain fruiting conditions.

Another method which you can use for the second flush, but is necessary for the 3rd and final flush is the dunk method. You're going to fill the tub up to the holes with tap water. You will need something to keep the substrate under water as it will float on top of the water.

Wooden dowels stuck through the holes in the tub across the substrate will accomplish this. Cut a dowel a little longer than the width of the Tub and set a weight on either side to hold them down. Another way is to lightly set something on top of the substrate to hold it under water.

- Let your substrate soak for 4-6 hours.
- Drain the water from your tub
- Reintroduce to Fruiting conditions and you will see a new **pinset** begin to form within the next week.

After your Third flush you will Notice your substrate is shriveled and much smaller than it once was. This is because the mycelium essentially ate all of coco coir and rye berries and what you are left with is mycelium and matter which was not compostable by the mycelium.

The substrate can now be discarded and the tub cleaned and reused. If you live in a tropical climate you can throw your spent substrates in your garden or create an outdoor bed for the mycelium to grow naturally in the wild.

INDEX

CO2- Carbon Dioxide

Coco Coir- A substrate made from coconut husks used for growing plants, packed with Nutrients usuaually sold in a dehydrated brick.

Colonize- The action of mycelium spreading through a substrate.

Desiccant- A chemical used to draw moisture from the air

Fresh Air Exchange- Natural air constantly flowing into your grow chamber. Mushrooms need natural air to fruit, not just pure oxygen. Natural air contains mostly Nitrogen.

Humidity - The percentage of gaseous water contained in the air

Hydrate - cause to absorb water.

Mycelium - the vegetative part of a fungus, consisting of a network of fine white filaments (hyphae)

Pasteurize - a process of partial sterilization, especially one involving heat treatment or irradiation. Removes 'bad' bacteria keeping beneficial bacteria.

Pin - A newly formed mushroom

Pinset - The distribution of pins across the substrate

O2 - Oxygen

Rye Berries- are the whole-grain form of rye with only the hull removed.

Spores- a minute, typically one-celled, reproductive unit capable of giving rise to a new individual without sexual fusion, characteristic of lower plants, fungi, and protozoans.

Spore Print - The method of collecting spores from a mushroom, typically by letting them fall onto paper or aluminium foil

Sterilize- make (something) free from bacteria or other living microorganisms.

Substrate- an underlying substance or layer, the surface or material on or from which an organism lives, grows, or obtains its nourishment.

Yield- the full amount of an agricultural or industrial product.

Definitions obtained from Oxford Dictionary

NOW GET TO GROWING!

www.ingramcontent.com/pod-product-compliance
Lightning Source LLC
Chambersburg PA
CBHW031940170526
45157CB00008B/3260